MW01287110

OUR HUMAN SHORES

Our Human Shores

Josh Fomon

Black Ocean
Boston · Chicago

Black Ocean
P.O. Box 52030
Boston, MA 02205
blackocean.org

ISBN: 978-1-965154-01-4

Library of Congress Control Number: 2025931944

Printed in Canada

FIRST EDITION

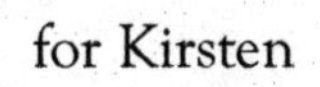
for Kirsten

Table of Contents

I.

Our Human Shores

‡

First the silence

then the break. The slow

roll of catastrophe. A country

reopened like a wound.
The rage below.

The love bestowed here
the decaying dream unwept

like two parallel timelines
of trying to hold you
again and always discovering
what it means
to be broken.

A cancerous growth. A kind
stroke. This ballot dropping

like a shallot into stew.
This witnessed strophe.

We take ghosts and whisper *infinity.*

Depart in carved-out oaks
a gnarled breath.

What comes together.
Sustenance held in bowls.
A sacrosanct cleansing.
This wave crashing
like a lip onto your lip

soft, sustained, I suck

your rift—that empty silo—
like oxtail, the meat

tender and dripping fat.
A misshapen consumption.

This oyster fresh and milky.
A purity, a purifier.
A briny life in parts.

We're always a slice
away from splitting.
The matrimony
you have staked.
Rippling red

waves over massive tides—
the day glows awake.

Mirrors our memory,
this taciturn break.

‡

Absence makes of me a skeletal transmission. A future silence hummed into the ether.

In a fog of salt, the marsh blesses the morning. Reflects all our history, our inhuman attempt at living, the sentimental idea of power. We see nothing until it reveals itself or we seek out how far we can bleed. Human instead of human, another caterwaul clearing.

But this warm morning we wash ourselves against decaying concrete slabs. We watch misshapen waves pour flotsam foam over a rippled tide. We seep bleach and outdo the limits placed on us like the water-soaked logs piling on dead streets.

A flapping door wags its life awake–I step through the frame, the threshold that once stood here, once saw the ocean at a distance–the water slams it back close. Slams it open toward heaven. This music we search out, this magic humans hoard, but always this slamming, this brackish song, this throttling praise. This new way of living.

‡

So easily we corrupt
        with uncontained contagions.

                In moments the blight
        beckons in moments
the carrier   ignites

        a plague–

                beyond the systems
                we believe we still
        die slowly.
                More readily

we rearrange pebbles in the sand.

We bask in sun
        this promise

        like a catastrophic trophy
                of living.

Our human shores
where we lap at the intake
the emission
of the operators' catatonic decay.

The want to need
to want
for remedy, a throbbing naked still.

A parlay of energy
all at once bright–
all at once
a shudder of bodies

flooding into loving
bodies, this tragic art
we tie down like waves.

‡

The threshold mounting.

This invisible line.

A warming, too   a bleeding night.

This light sliced right
between us.
Maliced callous   beached
onto me.   A hardened

fossil of who I once
knew. Time became

a burning catalyst. Denuded, riled
toward spirals of bodies

pressed tender against
mountains. We've never

named   our devils   called
them home.

This new religion

is a body count. A fading away–
a melting night.

Our ritual bleach.

‡

Quick collapse
secreted hoarse breathing.

What remains
after the break.
The tingling
sting of frenzied

turning
breach forever
painted
like the sky turning blue.
On the beach

we fire we flood we kill we cripple. We survive we leech.

The fog galloped
under each sternum
each rib   like an oyster
cracked just right.

A lingering still
—the whole
heart serrated.
This rhapsody bleeding
like a rainbow   spout
of rain.

You open the door

the wind blasts through

us like a sudden shiver
of loss. But what is gone?

Why are we still waiting?

‡

Such enthralling vigor, such prismatic velocity
at the moment of conception

it's like a comet consumed into a new
deleterious sway, new atmospheres of possibility.

We crumble before we burn but ignite the efficiency in which we smolder.

To have been in love, this is a story for you. Why do the motions create a vibration that casts a kind of shadow through our viscera? It's rocking fast the mind cast away, stripped like a taxidermied core—all bones to recreate the image and how swiftly it ceases to exist. In the morning light, you are a dream accelerated against time. We forget so readily because the mind creates a vision patterned just beyond reality. I have never remembered a dream in which I have flown. I have died many times and multiple deaths too. The shame and disappointment of loving too much and too freely. There are ways to shape the future, whenever we inhabit new kinds of places, new kinds of trajectories. Take what you learn and make it a sparkling mortality to savor. Dreams ignite a primordial terror we wait to come true.

In throes beyond the musicality–a dimension
in which our body exists
without our corporeal craving.

What remains is an idea
between the numbers zero and one–
a string of points that binds us
to the history of becoming
human beyond human means.

Like a body wholly body,

the constant pace of waves battering always the door.

‡

The riling chill obstructs for days, roils prayers to my father—a seizure from this numb brain. A sacred dumb landscape quaking with sinews and waves—in what? Its sacredness? Its landscape?

A mist stagnates like wreckage appearing before me—clogs the orange spirit of time.

‡

A tsunami extends
the order
of the ocean

inland, the mesmeric
swallow sudden–

enormity

blushes
like a kind

of thrashing
dance.

‡

Subduction zone of the uninvited. The quivering toil of life versus living–everything is contrived to cataclysmic change, tuned to an always state of shifting lives–between the true believers and those refusing to leave.

This passion played backward–a satanic verse hidden in the bobbing ferns. Diurnal somnambulations thumping your guide astray.

‡

Sublimating expression, that, in its own right, is the very face of god. The bare face of almost there, washing over my cool ruin. Positively the revolution. Gestures of the obscene. A flock torn down among the algae. If I could efface my grist, my accumulation carved into bones, if I could hand myself over to my total evisceration, what would remain? Agents of history pull the lighthouse away from the sea and foment an ineffable word spoken in the waves that tumbles through my limbs like ice. It's November, my vice screwed tight. Measured anguish, measured fright, that in my passion is a politics of external blight. I'm not prepared for this future. Plastic covers the shore. My body dissolves gracefully into the sea like a stout head. A washed-up jellyfish, a sacrament lost in wisdom, the seafoam frothing.

I stand at the shore
    dig deep
        for land. My skull fills

    with water ideas bestowed
therein. I'm warm now,
    fading always

before the swell.
Always spread
in sun.
Salt taking my
shape
making of me
a shore.
The inhuman.
The truly living.

And you, here. Salvation was never shorn, never granted.

‡

The petrichor bulging from within
the skin. The face turned
open to supplicate
public shores. Triptych we became

incongruous, messy
like a moist detail, newspaper beyond

salvage. The cheeks an access point. Where we bleed
apologetic. Apoplectic.

How do you part a new love into the folds
of yourself—of yourself
is this the imprint you want to lay

bare. Your body's crevice

impressed on my bed
a longing that can't   hold on.

It's been so long
in cacophony. Where have you

been parting the vestigial

yawp, a newness breached within yourself

like the moment you emerge
from your mother, the repose gladly mistaken

for stranger moments

like the full lust of life bellowing.

This vestigial life. This vestigial breeze.

It moves me so wholly. No, it refuses

a type of reflection
looking backward.

Listen to the echoes
reverberating back at us
from our waterfalls—it's seeking out

as much as looking in.

To step out into the night
a tattered genuflection.

It's an impossibly dark sky

a blank pallet longer than we can comprehend,
apologia moved through us like a prayer.
Language the only half

of understanding that is nothingness

rote into every breath.

Tell me I'm alone. Tell me I'm not alone.

That we glister enough
posture to denude slickly this sensation.
This reckoning
I reckon we should do something about it.

I'm the kind of blister
a band-aid won't help scab over.
A battered bastard we all learn to become.

A deep, thronging wallop
like yelling fury into a well—
a full-hewed echo

shaping curves of our mouths, sounds
the precipice ignites

where we morph into poignant earth.
Where we bless the surf.
Where we cease to persist.

Tonight, the stars won't flicker. The day won't strike
through a symphonic uprising

of rotation. There is only

you and tonight you'll pretend
to be your best self
shimmering

violent truth.

‡

Scattershot, a throng in massive orbit.

A trajectory you have claimed to have already heard.
Blastoma lining magic

within the straying fire—we can melt and already be degraded.

Ample self within another self and lo—
stubborn birthing stubble.

In the morning we can pretend

we are free.

Sizzling the fish, the trembling stare we care to repeat within moldering love. Heat is loss, a transfer of love that can never be reclaimed. A stake within a kiss. An unspoken bliss. Giving always a part of ourselves, filling it, and repeating endlessly the thought of how to never care what moors and sets our soil free. Do not remind me—I have already spoken. You en-

gorge your want, toss the piles.
Burl my tatters, fill me with
things that do not matter.
Plastic bubbles in a stubborn
boil. *En plein air* echoes this
catastrophic remedy. Death
prescribes, leaves stale brown
when the ground needs renew-
al. This septic care. This tearing
moon. I cannot honestly know
how long I will stay, but now,
enjoy the songs that molder—
an immediate silence when all
we build is washed away.

Me, here, always.

In the shore

we pray we can find the warmth we need

to believe in human

horizons. All can be intertwined.

All can drain in a day.

Let the darkness arrive and sing

macabre solutions—

winter will never be a part.

‡

Just before the tides surge–
a new coastal infatuation.
We lost our people.
And I began shaking
with the breath of time
—an oily feeling choked inward.
We waste our refusals
to do nothing while this fugue
of death makes us refugees–
strangers on our own shores.

‡

This extraction is imminent
I choose the mythos
we crave.

‡

But it all doesn't sit right—it never does.
This flight burdened in perspective
obscured while fighting other narratives known
and hidden. Together we try to make sense,
try to lilt words from our mouths. Slack-jawed,
replete. Its curious production. Its miseried silence.
We lift pages to create our coda. Suffer ghosts
in the milieu among friends. We sing loud our haunted presence
that years later feels absent, feels misused. We burying our beloved.
We piling our existence. We're the survivors,
the saviors of our own story—until we're not.
Until the rot makes sweet this fecundity,
barren, filing away the spurs of this infidelity,
jutted out from the breakage, the throbbing mystery
still. I drown in this hole that I've dug, replenish
this depth pictured like a star. We're torching
in death. A face pummeled, a gentrified rune.

‡

The end times are a braid
of veins bursting.
Humbled. Smooth.
Faded.
Wrapped in pieces,
our lives let slip
away, we marshal
this waste.
In the book
ink will pray
this catalog
that—at the precipice—
we pour out our memories
like a deluge.
My nostalgia
nestles like a pet,
a sickness—
like a world over
the world that
exists as menaced
energy. Always dying
into moonlight,
always righting
the silence
from mute memories
where we pray

we are the heroes,
where mouths extract
the exact moment
of salvation
where we breathe out enough
to breathe in
and take another life. Make
of permanence
this longing
this belonging
to something infinite–
a malleable plastic,
a fuming toxic ruse–
ideas today,
malaise tomorrow.
We crouch, our tempest
muses loud a new shade of blue,
a burnt misfit weather.
A distant
penitent memory
lapping forever
at your shore.

‡

The vistas drone into memory.
Drill through me like an imperceivable bridge, drop deep
the sounds buried in our longing—our exhausted silence.

How many epochs do I need
to form an ego? On me or against
me, every distance explores
the body's shores, the limits we find in sand.

On the shore, at the waves' denouement,
the ocean reclaims our sandy feet,
frigid brackish sluices between our toes.

In the moonlight, we envision another world
—a chimera between the unreality
we have forsaken to lengthen our lives.
Let go of the promises we keep dry inside

so that the phantasms explode into brilliant
sunlight for those whom we love
and are certain to lose. I let my body

wash over your body, the infinite lives we find within.

‡

Every day this face

limps toward collapse

—bows

to the apocrypha

dancing like fire.

Deepens its furrow—

when all the trees droop

and bend their weight overhead,

we will say we were meant

for leaving— that we even could.

‡

I used to have a name on this limnetic flesh.

But what does confidence inspire?

We're trying to survive an entropy morphed silent.

The excision chiseled onward.

Ice retreating over a smooth cliff.

Like a cold wavering dream.

Around this mist. Clouds fisting a snaky grey.

I enter ashen, pallored and mottled.

Embered transient, long-forgotten days.

To name the resistance in my throat
as it distends. Cricketing harps. Like
a trueborn lyre. Squeezing lavender
and rosemary from foreign homes–
thieving tender succulents
dark siloes shadowing
morning–stronger habitations
of brighter times–I inhale

and exhale the evacuation.
A bouquet of herbaceous fullness
like plucking stars
from a wounded night
and eating our rhapsody raw.
We make new this birth–
who we are gripped by the tide.

The tight, dying summer–raucous monstrosity–

is already cold. It's already ripe. The breeze defiant,

evacuating my briny slim body. Over sand, forgotten,

trodden flat, I'm weary, emboldened, welling up

to release my hands like joy, a bursting light.

‡

Peel the skin peel
        the pelt toward godhead
captured like the morning
        where we bled
                            out our sorrows
                  to plant new tides
                            the plastic sounds
                            spooling
        pulls of celestial want
                  in and out
pulls the body
        closer toward perfect suns
                  a majestic resonance
boisterous moist echo
        you are in my mouth
                  freeing me from parts
that promised a new vision
                  like corpses sprung free
from the core
        we are making kindness
                  a pattern
                            of our guilt
the whole version
        where you stood before
                  me and knifed at your heart
                            pulled the whole beating

out and put in its place
a misery we shared
and here I am again
lost among the stumbling
bodies feeling a semblance
of each body and feeling
each other up and even
the dead ones cast upon
us a memory
where bodies
used to be bodies
moving through the pardon
us and pardon you
too we are in arms
and armed we destroy so easily
the kind of visions of ourselves
we'd like to be and be in
stolid fugues like a whisper
all patterning an excuse
to exist in this space together.

‡

The moment when brackish ocean
transmutes fallen tears,
assimilates its waves–it's raining
today, I'm the grey hue
puddled in streets
reflecting back why the sky
which itself is deep white
hums an incantation, the permanent
mood, an iridescent stain
made permanent the swirling
dune.

‡

Palpable grim, grey
bore into grain
siloes. Revocation of silent
roars, its distance nigh
—no, it was an aural excavation—
our world on fire,
ash shrouds the dilapidated
whale bone—no, Steller's jay
nesting—hellebore among
the charred logs—what is caved,
what is wide, stolen—a territorial
revelry. Cackled ash, incremental,
new heat deepening. Festoon
of comradery, a macerated
silence, ice breathed in
sad, relicked haze—no,
bellow a revolting prayer—
no, begin a revolution
within your hands,
froth furious a new life
for all of us—no, it'll take
everything we've got—
it's everything we've
already taken—

‡

Love is a musical incision
ignored willfully like an epilogue
about to turn
into a new beginning
an acknowledgement
skillfully mulled
enough to name that which we shun
mutely but not enough to pirouette
a sinful art–we are
struggling in the corners
where we pull ourselves apart
to break nature
and feel life thump open
like an aria
of gulls circling
a crescendo of waves
going silent.

‡

In storms my heart cracks
for you
in explosions
that contain
all that will
be.

‡

Scattershot, a song in massive drift.

A diagnosis you have maimed, have listlessly merged.
Lust miming tragic

the flaming sadness—we say we will be retrograde.

Pamphlet your self, make sure you believe it. And sow, together
sizzling gab, the misspoken, the reclamation.

In the morning, we send free

the missive, the permission to land—

All will be forgiven. All
have been retried. In the
morning, the sky lessens
what can be forlorn—that we
raise together an action to
flee from bed, and then—we
fill ourselves with darkness
like an ignited sun. First,
the gulp, first the throng, a
small passion hummed like
music, a chaos so forward,
so momentous, it sings a

kind of creation. A martyr
wounded martyr into what
we could be.

—and then wisteria.

It's in the breath, woe,

that we glisten, reflect the promise

of every day into a fissure. This purple loom
falling off our limbs.

Depart from the exhaust, energy held within. Explode
this impossible human flare.

We can bloom together, silent, as feverishly
as we ring like bells—

You and I will never be apart.
I sing with a mission, a knot into a missive.
Send me your fox.
Send me your fox.
This human threshold we acid wash.

II.

The Memory Machine

‡

Born I saw the contemplative blue
cracking, greying our refuse
our refusal to move   my country
a stale precipice.   This whole smoke sideways.

Inhale—now—that breath.

That erogenous zone
of order—hotness exhaled while we wander.

Our pilgrimage   to parasite new opulence.

My skin never used
to glisten never used
to sob the leftover

parts—glances, fears, the unending mirrors—
we tossed all night   on the moon.

This blistering cistern
this sweet rotten
shore.   The dalliance
of you and me and you and me and you
and the overtaking.

The capture of slow pleasure
warmed–emulsified like fat finessed from meat
the buildup and refractions. God
gone to the corners
of our hearts.

Save grace for other facile
icicles last recorded in freak
discovery
we cheer our awfulness we've prayed.
It's a decision, an indecision

a hot hissing
silence congealed.

Let this be the blessing
the slit in which we see
god in ourselves–

I remember eating stars, touching five fires all at once. I transcended histories within hibernal roots, macerated vernal hymns. And you–you were perfect. Everything wrapped in my limbs. A long drought stare. From the window you found me and skewed me bare. A wallop echoing outward like mildew in neglected corners, vines creeping slowly on for-

gotten logs. A putrid fallow memory,
my spine split like hands unfurling
prayers.

The buffed-out metals
rusted through.
A belched-out mantra—

It's me. It's me. It's free.

Pressed new conflagration within the sorrow
we wait, we miss that which we don't know is missing.

‡

Bursting inside the seasons there were moistures and colors and vibrating browns. Oranges engulf the shape of things. At the pinnacle, spaces where magma suddenly becomes a floe. Pulchritude greens into verdant perfumes. Azures beyond the comprehensible sea. But then the barnacles manacled our journey, demanded their blood rite with their corpses. Ghosts bled into the edges where we write our ending.

What exists in solitude cannot be forgiven cannot be touched—my lesions were blessed—caressed certain. Larch leaves shed, leaving bare the yellow waiting to smoke another tribulation. A sweet grey of stardust—enervated rage. Chthonic, shiny death.

‡

Burgeoning into a thousand folds of new selves I forget so easily what I think I want—years molding creation. Faces that pant. Solace in kind, measured distance. Even deterred, there are words in our isolation for the harmonies—birds we mispronounce, the constant pitch ringing. Don't be mistaken, I'm a vast broken thing I want to piece together. The traveler stepping into a new city—derelict in possibility—we forge on until we become incessant. More whole. I want to sever the blood stained over the world I love so wholly—a thick lacquer dripping immaculate humming from the strange cavernous humans we have become. Like a painter's grip we flick a decisive line that will change everything—then the chemical wash—grip me so tender I forget everything in your stare. Becoming a plentiful cleanse, I reinvent my misery, a ritual rebirth.

‡

Etched

toward ruin—the Book of Inhuman

Heartache—on its own—

plows toward the apocalypse. Deepens

its lineage, its cleave.

Imprints a mighty collapse like age

rushed through me all at once.

How do we mourn that which has
yet to be lost? Awaits a horrific end?
Gloom is a despondency we can learn
to embrace. Forge within our ferocious
survivorship. We can love the flora and
the fauna, we can cultivate a resistance—
in time, we pullulate utopia. The epoch
we wait to dismantle.

I've gathered the words, crept them into
form, patiently awaiting their demise,
like dying wood. I build
the house
you enter.
The salt air, the sun
bleached over this rising
tide. I've been watching

the sea seep into the book
and slip carelessly backward

with wine. My head opens
like roots overtaking

this captured malignancy
growing feverish, isolates
our preservation.   An archive

all over—before and
after—a layer of synthetic excess.

It's profound.
It's simple
and perverted—we were always gulping
water, always digging our heels

deep into the pastiche
the marauding silence.
When beauty
captures our tumor,   the sweet
unknowable death, we kiss, we shoulder
past our fervor—

we move on, carry our waste
between us
fading distances into intention—
flickers we pretend to touch.

‡

Buildings meant to last—shadows of empire upon us.
Stoic relics of conquerors and the conquered.

And where are we in this stolen land?
Destruction, decay, everything made

to be broken. Not devoting enough innovation
to even supply parts and the well-worn hands

that can fix our history. A sad spoke
tossed out instead. This isn't

the warm, crisping crust of an apple pie—
it's exactly where my thoughts breathe.

A tall dereliction, a meandering quaking incision.
My clockmaker fixes an irregular rhythm within me—

the precipice of emptiness like a spider
spinning a special kind of web, so tiny, so duteous,

time would exact revenge. Truthfully, my skin's peeling
like an empty house in the sun—we put coats on ourselves,

but will never be tough enough to stand when wind
uncoils listlessly the din silence of dead windows,

dead porches, dead stoops, dead dens, dead kitchen islands
cycling waves of rain drowning a deft numb into nature.

We crash against the sand, wash ourselves in the big waves
overtaking our bodies, tongues pucker like the desert

collapsing, a cold so full it swims through our bones
and pulsates aching muscles we have saved

for a moment of salvation. The drop of exacting
toil, over and again, we take the water in

until the day's sun creates an outpouring,
a gold messenger welling up a briny grin.

Dazed until I feel you grip my hand to pull me toward shore–
lo, soft upon my flesh, there is another flesh,

like ablution, we are all wrought and pining perfect futures.
But the day I left my city, the world gulped

in spite. Tore through me a sad shell shorn in two.
Cockled the blood from my feet. We will never return

to this time, there is no retreat–you must change your strife.

‡

The weight of the world
exhaling exhausted air
we rote into every breath

Tomorrow, we have shadows
to earn–lies to fulminate
this ignition, excelsior our loves

This breakneck maneuver
moves us to the stage–
a new position in dying

Lover, you were in my heart
perfuming the smell of history–
every great love begins with persimmons

‡

It's a precise mantra,
musical misprision,

an innovation of hands.
This skin made new, dolloped

like a glacier. Stoic repetition.
A cleansing missive

of too much oil held within.
A toxic seepage in time.

Muscled parentage of time.
The extraction skillfully exacted

—a decommissioning of body
mulled through restraining

expression—a holy pervasive
thought. Here because we choose

each limb—derision pulled like a wave
over the corner of each eye, held

like pregnant breath—the parts we break off
from our wreckage, our jetsam jettisoned.

There my body is invested in
other bodies silently reconnoitering

new hearts, new burdens to bear–
silently naming that which we spindle

mutely. We have begun to caste
beyond our casual obsessions and now

we need to make the rich
ashamed of their richness. Acknowledge

capital silhouetted against degradations,
becoming impossible to ignore an ignominious discomfort.

So, discomfit your shadows and untie them from the ground.
We are charged with gravity

with clear, moist skin. United, begin each day
with a question of how to subvert

the consciousness we collect within
a blistering foraged heart.

If you erase the snow off the mountains,
promise me you'll repudiate

that contraption, your face.
Use a taut line of making new

that which cannot be renewed.
I–mulled over–witness to

sporadic obsession–a wholly pleated
person folding onto myself until

you–pulsating anomaly–
burgeon together like the winter run.

‡

My sweet catastrophe–
a cascading cool summer lemonade
spills into a plague of ants.
Watching them crawl on our legs,
I had the thought we're always dying,
always undoing what we love
to rebuild resplendent joy–
into the moonlight, we quench lewd.
We drip the harbor like bloodletting,
a tidal liturgy of letting go
and holding onto something dear.
An undertow eddying our escape
toward the current. From the ocean,
the shore drifts away.
I'm all alone, the night pressing
its humid touch like a sponge.
Forgive me this original sin–
existing for myself
and forgetting you.
We're all guilty. We're
falling from grace
like an infection–
a lover's dulcitude.

‡

this throttled glottis
this financial ruin

whipped like a new savior
we shuttered our maws

made a pagan promise
nestled in quicksand

‡

I've gutted these words
placed them neatly

and precisely

into this tome–

in this hidden space–
we're patient, dying,

the salt-air succumbing to the sun
rising when you're near.

For the sailors
as they chart our seas, we

leave a pickled fetish.

A frozen fog
captures the exactness

of our existence just before the sun

burns off
the claustrophobic clouds
into the horizon.

This chronicle of calm

and disaster—bones folding inward
against bones
cracking, screeching
the soot
outwards. Toiled crumbling breath.

Make of me a fervored gulp—cough a hurried mile.

Make me listless water.
Always, we wrap into a pleasant
violence. Another hole deeper than the first.

Why can't I scatter our cause toward catastrophe?
This Aries
life is a restless riot.

A fire always burning. The bones
of my home are rattling, they're

barren, empty of you, charred—the echo
of our lives shatters

the back window—a quiet shriek
then a twinkling silence.
I don't recognize myself

filling the pane as the moon
grows larger in its infinite caress
toward darkness.
The tiny pieces and expectant
faces growing beyond

what I can recover–
a field blossoming
hope of another tomorrow.

Some things are worth saving
other things hardly exist.

It's in the books
spilled with ink.

Across the seas, we grow
larger, increasingly incomprehensible.

We become fully realized

like islands emerging
from desolate ocean–
a complete eruption

toward lives we can't
choose but inhabit
lustful chasing–a new life
we toil endlessly to build

together. Today, I want you

from an incomprehensible distance.

The ache, the yearn, the incomplete feeling
nakedness mirrors. As the sun

rises, I'll wait for you to bare
the horizon, blur the dust in between.

‡

We measure
        our intake
                a deep sagging
        bellowed our lubricity loud.
                We dreamt
the distance. All over an alacrity
        despite the hollow blur
                I must be silhouetted by love,
        by the strength you chisel away.
                And you here like a statue–
against this future, we tailor
        a bespoke love, nooks sewn in
                calamity.
This tropical breeze stirs my bones to extant lands.
        I'm still learning to love
                in this annihilation,
                this new day.
I crease warmth just for you.
                Crater my vastness
        into a breath
of the moon's spore.
                There's nothing but mountains–
                nothing left to storm.

‡

Sand consumed my home.
Desire torched my youth. A shriveling
made human this purpose
to belong. A space between
eras. Sounds whispered
into the roots of men.
A sad kissing booth.
An epic for what reflects in us.

Then a corpse keeled over.
This question of use.
Like water, we gravitate
over peaks
and lap into pits. Pool our dreams
together, overlap that snug comfort,
that fire bespoke at the shore.

We interlope, overtwine
songs breathing life
between small geometries
in bending blades
wind atrophies a melody—two
separate songs screeching
sweet cacophony.
Mellowing each other's
hearts. When you sang,

the melody pulsated sweet
benediction, a remonstrance
remembered. Like a kite
we sped toward the sky
tied our youth to the ground
on the chance we could climb
upward and sail free.

Promise me another day. Promise me
you'll carve a feminine infinity,
a statuette to bless the acid rain.
Tamped sporadic, it will fade,
eventually, like a blossomed
thought. I'll die like a flower
propagating our lives
onto everything we hold dear.

‡

BE I AND BAD BRINES

HOLLYWOOD SO AFRAID OF IT.

The kind of new motion of mouthing through the motions.

SHIFTED THE WORLD'S GAZE TO IT.

Still won't change a thing.

DON'T BURN TREES TO FIGHT IT—LET THEM GROW.

Like a sinkhole, we are told to wallow.

WARN YOU OF 'GIGANTIC' IT.

Because we call it a horde, we expose a killing, its raw tendrilled frame, its long tall tongue.

IT: MORE DAMAGE THAN WE THOUGHT.

We did not take to think, could not comprehend immensity.

WHY DOOM AND GLOOM WON'T HELP US FIGHT IT.

Why we can't even say it and the rest of its pallor.

THEY WON'T STAND FOR INACTION ON IT.

But we do.

IT IS NOT A FIX FOR IT.

Watered down, I became.

DEPENDS ON YOUR RICHNESS.

Extradition of the murder, passing the buck with our hands in mating season.

IT IS SHAPING LIFE CHOICES.

Yet life choices shape it.

IT IS A NEW DISORDER.

Today, I learned how to mouth my reflection on a mirror.

AN INACTION THEREIN. WORLD GRAVELY UNPREPARED FOR IT.

A room lush with efflorescing lilies, a predilection for flowers.

YOUR BRAIN CAN'T PROCESS IT.

But we keep making out with the mirror.

THEY'RE TALKING ABOUT IT LIKE NEVER BEFORE.

We never stop talking, never stop to listen to silence.

WE NEED TO TALK ABOUT IT.

DEFT, IN YOUR AMBIGUITY.

UNTOLD BENEFITS OF IT.

That post-pregnancy glow, that boarded-up ship, a clown out of water.

NO, IT WILL NOT END THE WORLD.

I will end you.

WHAT'S YOUR IT. THE CITY WANTS TO KNOW.

Ball gags and fist chains, a tenderness so queer, I want to make you quake.

YOUR ROLE IN COMBATTING IT.

Your inscription writ into my bones.

IT STILL THREATENS.

How empty have we become.

IT. IT PT. 2.

The sequel of apocalypse. A sheared-bare promise.

THE ANIMALS THAT WILL SURVIVE IT.

And in pyres, I will be with you, here or otherwise burnt into your skin.

IT IS MAKING IT UNBEARABLE.

We remake the world into an equation of seeing beyond.

IT IS MAKING IT MORE DANGEROUS.

So am I. So am I.

‡

Don’t tell me what writing is, don’t tell me it needs
to be mapped out. What brims this vivacity
is unspeakable, immovable yearning

pulled through a vision we are so eager to dream.
This half-measured breath blessed toward nothingness.
These plenary derisions we thump bare–

a mastery of life. I began to reject that which denied me.
I was a body buried when my enemy
found me. Light treason blossomed blue, the kind

bosom of borrowed time. I drank
the blushing light when they came to take me.
How alone we were to call

off the war. How close we were to eradication.
Sometimes, I want death to pass us
like a lightning strike hitting the house next door.

We fingered our pockets, paid to end this night
with lint and string–a whorl of shiny coins
that burst into life when scattered aside.

The snow romances wry this epiphany.
At the firmament of the woods
you said when we begin to die we pulse

our pretense away. The flush mounting
seasons we mount each other and hold on tight
to this violence we pretend to make. I want

and therefore am so lonely. A rife tone
between living and almost dying. This precise
wash over my eyes—there was always life—growing

in me and a life growing from me and there was music
scantily off like our nude salivation—a measured undulation
under my sick masculine duty, my rough embrace.

Hold me closely, open my chrysalis, blight my possibility.
Burst free this aurora, this twinkling shimmer.
Make of me your horde. Calm the violent air, the sticky

image of the dunes and the shore.
And yet. Here I glister. Here
I'm ready to fight. You were never alone,

never unburdened of me, my sanity, the deer
bounding at a sound's fright.

‡

In darkness crows cackle
and gloom overhead
plotting always

our volatile patience
we feast. Night wallops

a decadent end.

We've fought
the culmination
so long for nothing.

Another sunset, another
sun quake.

Grasp me tender
grasp me bare.

Inside me moved.
Inside me wept.

The pageant was upon us
and you were death.

‡

I see you polluting your complexion.
An untethered corpuscular gnomic trend.

Cleanse your house with this tiny dollop.
Age away your anxiety–there's enough

to despair while we repair our history.
Render the waste from our skin.

Undeterred, we cast aside a rippling–
the pain during the refrain and then the rondo.

‡

It's disgusting and imitates trash.

Succession needling

a memory obscured.

That lingering feeling of fucking a friend.

What is remembered but never celebrated:

inhuman strata, a prodigy, a membrane?

The very way we became human.

This pile of ourselves, memories poured sooth.

A cruel cyborg dream of what I remember:

The vertigo of reaching back

into history to place yourself

in a moment. Transgressive, misremembered–

celestial the horizon quaking our human excess.

†

Post-bear, classical green. Remember me

together, we're lean, full-fleshed, and weaned.

A fashioned tragedy forever maintained–

the promise of innovation, and then dying too late.

A prayer borne of bones–

and then the radio–a song

you'd have loved bursting through.

‡

This is because I face uncertainty wholly—a glottal reciprocity —a swelling heart inside my voice I cannot sustain. It chokes me like a python. My throat lubricates gaunt swallows, trickling remorse where I once had obsessions. It's a benign partner in how we approach the world—just enough sensation to know we are being strangled, but not enough to taste it on the way down.

I don't want to live in this world, at least not at this latitude—ice forms on my mustache in an instant. My mind fogs like the marshy wilderness. Categorically unfit for the shape of me, I descend. Into what is swimming in just enough to feel cold blown into my rib, holding me at a distance. Lowly capital. Baptism in heavy water.

I'm here in this colossal mediocrity. Spelling my own name. My own disaster. It's yours. Ours. A dream made certain into astringent augury. I'm certain we're here trudging through the mist. That in the morning I'll have reached you.

I write eclectic visage. A prize for the end of the world. Bound, bagged, wedged into your fissures, let me swallow this tragedy completely—the waves lacquered, the floods mounting, this hirsute way of living.

My enemy spreads me like a plague. Here, now, an infinity raged.

Chemicals entrenched in biospheres, microplastics at every reach. Technofossils of the dam's breach. We said we were waiting. *For what?* An impenetrable memory.

‡

Find me god in obsolescence, a crooked
din clanged like a rusting echo. The morning

pulls all the parts of myself together
I don't want to be pulled together.
Torn apart, washed away.
The methodical

sounds our body chooses to silence—
its fantastical streamlining of order,

this chaos we have sown.

Stimuli drowning us always
patterning that which deneuters

the sun's swinging praise. Meticulous this,
meticulous that—we must fly
when we dry ourselves of the filthy
liminal film after we wade into human shores.
Smeared, caustic sway—
a prayer blooming into a newly flayed day.

This winding, crescendoing silence.
Drone on me some more, pull this partition apart.

I come at the very end where the
roar drains itself and exhales too
boisterous a frenzied malice. That
in myself exists an order seeking
to undo all those around me, to
know each person for who they
see in the mirror. A religion of
me stowed into their pockets. The
self-care, self-repair washed over.
Let me save you from yourself and
me—a mystic dredge excavating
ruins. Miscreant, my phenomena,
my monopolized—whisper it, your-
self and everything. Blister in. Pray.
Bray. Be mine.

Let the mantra lay freely this violence.
This sense of wonderment
taken and frayed toward pasts
irrevocable. *Whiteness is nothing.*
Absent everything. Let it all slip aside.
Resist yourself. Sway the haptic into tingling penance.
Reclamation of the void. Reclamation of your voice.
In everything, this prayer is to be everything.
Understand the menace. Understand your measured
self in this undying breath. You are a place.
Your death is a trajectory that needs to be nurtured.

Bloat to explosion this excess.
Sew from sickened flesh, a kind
person willing to stand

always through raucous cacophony and
save those drowning and need to breathe.

‡

Weary

we chase belief.
Buried

in deep. We

exhale when
sun exhausts
still.

‡

In February, we marked
our bodies in stones palmed from the sea.
We dredged our life together
like a stone dragged into the coastal swell–
ebbing outward and thirsting inward
like a heart. I pang like winds pulling
the coasts apart.

In March, we became an echo–slung
against this house–
reclaimed by sand. I am learning
our bodily limits, the margins
our bodies limn in between
this empty space–
our new faltering staccato.
Yet, we forged on, we began
the family we lost, the span
of an albatross could not
unfold between us. A marker where
bodies lain strewn
in sheets, loving long
renewed.

Like a mighty fine grind.
Like a dowel carved inward
to form a table's leg.
Decoration and kind, strong limbs
contained us.

In April, you kept finding stars
on your skin, buried in time,
unearthed haptic rays
and dreamt of tomorrow.
The sun fused us together–
A new trickle in morning's cast-iron urge.
A body awash, a body in us.
How we exist on land.

III.

Book of Skeletal Transmissions

‡

And here I begin, the sharp howl of winter's immaculate grey. Wasting away, now, in the throes of summer's last romp, the burgeoning long days kept in check with a dumb sense of wonder. Call it a blowout, a bellicose brain. Too much to cherish, too much to resist. To perish, an option, but we'd rather not.

What do we become in the numb numbers bellowing toward a countdown of too soon? Stars blow up until they are born. Planets must cool enough to create. The brain is a deficiency of knowledge. And knowledge a deficiency in access to forms we can mold our bodies. A turgid kiss. A faint recoil. The winter tiptoeing its creep, making erect my bristle.

Whenever I'm alone I call out "Hello-o-o-o-o" to affirm my existence echoed back at me. That hard silence. I want to cancel my affliction, turn everything off. God is shrill. Throbbing cysts. A growth hemorrhaging silence. A festered open wound—muscled, pulpy flesh. The tendrilled undergrowth of a dilapidated epoch—this fidget spinner tick tock. What is the moment when everything blacks out? Where do we inflict its song? An echolocation beyond where we want to want. Where the hum howls its soul to break.

This torrid parlance. This understated rage. The voided knowledge they say when my skin is burning.

My enemy is covered in tumors. He screams daily at my door. Stands inches away and wallops so loudly the door shakes a perfect resonance and vibrates a solitary hope awake. A simple lock opening, screeches like whales—so many whales—stranded on the beach.

Hello-o-o-o-o!

‡

What drives us mad with desire
to topple power over the null.

In between the whole. A never
completeness–never architecture

of full. Steepled faith in nothing
keeps me going strong. Entrenched

silence echoes death before we sing
with a beautiful final breath.

What will you say when your body
crumples itself to the ground? What

marks will you have earned to say
you really mean it? Our birthmarks

are how we show children we have
debts we cannot repay. A primordial

threshold toward hope. Always finding
new ways to truly mean it.

‡

You imagined normalcy as some declivity in your heart, something you always had but could never reach. It was like growing up with an ocean outside of your bedroom window that disappeared when you looked at it. A pleat of sand always scratching your heart. Infecting everything you touched. Waves crashing but never near.

When your enemy made me his enemy and together we fought, we had sin, we had not. This litany. This pit. Within the cherry. It made me whole. Made of me. An incurable sickness. An unfurrowed regret. It took so long, you and me, to realize we had been defeated. That our enemy was already there—had been there burrowing—before the war broke our whims. We were losing always these unthinkable lives.

In the streets, I scream for my bastard to come out. But he was there—permanent like a spout—fisting the ground. Making of air a motion, a thick, juicy tear in the possibility, that years later, I wouldn't care.

After the war ended. I wandered. Pondering so loudly my feet began to bleat staccato repetition. I bled my veins like the falls, the water a kind of joke, an impossible line to follow. A nation filled with seeping bleakness. That tangy bland.

We began to molder. Began to quake red. This cistern wetting inside my soul was the little resistance I throbbed my life toward. Until the moment exploded, I held my breath to face the blow I knew would break away. I fled right to where my enemy practiced jaundice. In the mirror, I rippled.

My enemy was around us. Misshapen. Wailing like hounds in wind. Throwing himself everywhere and all the blood and fire. A reckoning we pretended we weren't quite near.

A misty cataclysm. Caustic god where are you. And yet.

‡

The signs were
there. Over air.
Waiting

to be fed.

‡

Hundreds of tethered objects flapping in the wind like only bound objects can. A whipped murmuration in life's simple chaos—a void replacing another void, endlessly, violently, not nearly human enough.

When we replace one silo, we outline and inhabit the rims of our want—another silhouette of our hearts panging. The whole bleeding lot, the fracturing, smattered ends. Refuse and heartbreak from which we breed creation. Piling on our lives to new lives, creating a simulacrum that we must parse meaning, burgeoning reconciliation with new from old. This maddening shattering. This touching distance. The pile where I emerge and never look back. These human shores.

Tell me, what do you excavate from your memory? Where were you when you loved it? How does your love reveal all that you want to share? What footprints do you leave behind? What songs have you written? Is it enough?

‡

There is profligacy in the repetition. A cooling heart tossed on the street. I lurk at the jamb, the door opening its maw, trapping shut its subordination. I need to know what becomes of us when we leave.

I had this thought in a time of annihilation—the opening season beginning to hotly rot. A flourish of tympany and subtle grandeur—whole worlds crashing into other worlds. A gnawing, idle still.

Trapped within, I reconstruct a season exhausting—a malleus oscillating to shrill, dark bends, porous juicy fog. My enemy crafts darkness within you—blends encroaching pangs to show contentment is a drop of terror. The days grow shorter, the ravens nip at the canyon's portentous rim.

That which is unknown sows fecund looms without a stir. Causes histories to burnish.

An imperfect, radiating mass like an inoperable tumor behind the thick parts of my skull. I long for the incision, the barrier of unknowing so near.

Lacquered, like an object obscured, I thicken my ice bloom.

Fall implies want, intent within error, passage into unknown apocryphal forests.

And then the terror–I'm afraid I don't know if I am made of that which defines me–or that which destroys me.

‡

and of love
        there is sickness
        flapping

‡

The book is an unwritten infidelity–a lie that anything can be complete–a benediction toward an infinite incompleteness written taut, gritted raw–a primal howl of caring enough to carry on to new horizons–the books beyond the books, a patient penitent scraping. A wave lapping at my door like a perfect, discrete pathology of love–a lover recalcitrant to losing faith in the bigger version of ourselves we place in each other. We reflect in the mirror those who we trust enough to scrub out our patinas.

‡

This curdling gut, this barren ferocity is every ruthless intonation leaning in like a reverie. A kind of plaque plastered onto our veins. A real block of misery.

I began to believe in the chorus. The pain cried in unison. A soft, eerie singing plight, the new fervor in our survival. This apparatus we glissade.

I found you, you found me. We were buoyed, bound of our bondage. Life played out in liturgy. Wounds effloresced like potpourri.

I place your leg against the cliff, your wrist on my cuff, our thrashing together became a kind symmetry we curve in each other's faces. The thrust of your coast won't fade away.

The tide ebullient beneath us, pulling us closer still. When I looked, you weren't there but I knew you'd always be nearby. My infinity everlasting.

That's why I sing—I sing, I sing!—into the night this elegy—a turbulent froth churning awake centuries of raucous sediment, parallel waves bound to forever crash alone and never complete this missionary—a mourning slung into low tide, a scandal about to wake.

‡

Fatigue me in parts

let gold grip shrewd

these new tatters

glinting the night's

brazen glaze.

‡

It's not quite what you're thinking. Let me begin by saying I've been made whole. Not reborn. Not my new aging dad bod. It's fond recollection, wholeness found from the memories my muscles ached real, the world quaking through my reign. Victor, my distant twig, love is what I derive for you in changing my detritus, new visions—pray for our salvation. A comet marks a generation, a mass extinction—verily, we chase. Blurry we paste missives. Dead men blister the edges. It's past time to give new life to the book. Those skewering stakes to embrace the fall. Those birthing a reckoning. A body within the body altogether new.

It's only a matter of time before the blueberries stain our skin like galaxies.

‡

At night, I retreat into anonymity, surround myself with other selves and strangers who make love to my sorrow.

I'm only learning loneliness. That even as I turn to create, I isolate my annihilation, I chameleon wonderment and awe. Ubiquity beyond ubiquity. So easy to fall into delusion. Comforted kind. A hallucinatory brine.

I night for power. Nightly, I cower.

From isolation, I draw out my veins, the scars I've earned. Recollect that which is not a part of me, that which is always apart. Night an empath, a quiet reckoning. History stitched together and severed and re-stitched. Like spreading fire, a ritual, a violent rite. At night, I'm whole, a burst echoing in the atmosphere.

The emptiness we ascribe beleaguers possibility. Shines opportune narrative, that in ourselves, we collapse into poignant, mundane heroes. Right now, I'm pointing at you. I'm staring directly into your eyes. My threatening enormity. My bland countenance licking your neck. Virtuosic quivering. So much exasperated air collapsing out of our bodies. And still. We persist. We fight. At night we break our bodies like crumbling mud.

Charred, the red added later–deny our death until the very end, calm the warming pleasure of our human weave we entangle early.

Stillborn. And yet. A morning sliced clean. A morning missed. We night its moment of action. We write the body all wound together, knotted limbs collapsing on sight. We write the history to construct over decimation's residuum. A marker of where our bodies existed and manipulated time and space. We write this pleasure, this sight. The feat in anxiety that parts us. A new dew hung on air until noon.

Will loneliness become wisdom? For years, I've missed my friends. The ideas that blast from their brains, their colors and bursting visions. I'm a moon collapsing in on itself. A tide consuming against the breach. I'll guide you in this wasteland, this blistering beach.

Like a body against body, this melancholy sweet.

‡

Reckoner, will the rescue come like a sough or crack the tree in two? This dilemma questions its footing, meaning something *you want to have meaning*, but undermines your roots. The moment we face loss, our regrets escape the high-water mark and overtake the parts we allow to be loved. Admit you are uncomfortable, let go of your rope. Focus on moving parts integral to connect our lives into a hirsute apostrophe. I'm not here. I'm not telling you what to do. And yet–

‡

I choke on my fetid heart. My clog of mythic order. This belligerent maxim.

It's not enough to plead, not enough to make that which is wrong, right, or slightly more. Like a fox pouncing on mice, we become too deliberate, too afraid of life.

I peel my ventricles meticulously. Bleed my promise to come home into tomorrow.

‡

Electric waves wash over—a sea of songs
fading infinite the disaster we clear from our throat.

‡

I would like to repeat myself, no, the martyr I have made in my memory, the aching body, ghosts never cease to linger after cold, primal encounters.

It follows close, this liminal season—I muddled myself this way. How, the incessant snarl of my mouth caresses a kind of hum emitted from my teeth. When I was twenty-three I committed myself. Into asylum, into mystery. Barking soles. Sweet, barren flesh.

It's the affliction, you staring there, that we can't diagnose, the words harmoniously seeping. Don't stop plastering my vision, don't stop the recursive anxiety. When everything threatens to implode, I lay here, let my poverty strangle hope—it's what matters, what sharpens from the luxury of ideation. A failure of licking that which sustains me. Reversing a gulp. Receding from shores made inhuman.

How can we maintain distance when we don't even understand where we stand? The madronas glisten best when they're wet.

Affection. What I pile up on the outskirts of every city. It's mucous on me, that simple permission to sing. It's a lonely poetry of moving what melts into stasis. I apologize for all the death in becoming who I am. That spew from my–the conduit of everything, that body folded over–and then rippling together, blends a new entry that pulls my tide and yours. That human shore, that tender-taken breath, that blessing dressed in tide–or else we swoon to death.

That boy could never have known what became him, would become us–so glorious! O our pollution. O our patient misery. Hung our sky into blessing, hung our sky like gristled teeth. That snow a distant putrefaction. That melting concern of all we messed. A putrid sterility.

My human shore pours through my glottis. Pulls everything through me, pulls my guts into my hands. It's pure art–lonely, right, and rigid–but I share it with you, share the rot, share the folds of my fat, share this moment. That here, now, we chant, you listen, we chant, you glow. Glistering this viscera we wring. The lives we bring and unfold. They clot and squish, wash slowly ashore. It's all I've got. This glisten. It's all I have to give. All I had to bring.

And you here. My constant breath. My absent humming lingering toward death.

‡

We cast away the shore. Unbuckle our predation. We speed inward—in land we burrowed like a charge, bolts screwed to my thigh, that time we hiked into the sky. This thistling memory—a week later we forgot our boundaries and streets—I loosened the city's limits. I dream of Negroni's bliss, the summer heat lingering discreetly like fresh rain. I wake to fire, crackling, roaring hinterlands. The season's lost they say. Another pox upon the blooming frost. A new unending tide. We cradle our moors, take from memory this future promise, this indigestion that doesn't sit right for years. In time, our friends will all arrive like they had never left, singing out *Where was the boogie tonight? Where was the boogie tonight?*

IV.

The Somnambulist's Lullaby

‡

Alone on this coast

Is there a ghost

Making blue this ablution

The detritus turned chatter

Lingering like an oil spill

Ebbing flotsam of who we want to be

‡

I strip bare the marrowing cold.
I stand whipped against the ocean

the cold snapped sand pelting
a precise explosion wasted

into distilled truth—an incipient
thought flowers—pollinates

a facsimile of fashioned hours. Hope we make.
Hope we need and always a burrowing.

Here, I collect the chemicals
from the air. A frosted, wet morning—

exhale these vapored thoughts
extinguished immediately.

I try to drown out the waves
that are drowning me.

Echoes shape the edges
of memories we bleed

together. Emotions we forget
to forget. Love emblazoned upon skin.

Call me a wreck. Call me when I pass fidgeting
silent into pulsating Ragnarök.

Promise me tomorrow.
Call empty my name, a broken

breath of me, my memories
you have lain.

‡

Timber to the apocalypse

the dilemma clear—

burn everything
burn anything near.

Sewn into my martyr

words become
kindling for the end
of the world.

I lumber up the mountain.

Air thin like gasoline.

We become
new fervor among
dying sentinels, a spark
toward humanity.
Listen clear—

When forests ignite they
roar beyond comprehension,
beyond boneyard night.

Beyond the sentence's capacity,
a fresh toxic missive blooms—a
colony of crisis, land beyond
borders, stacks of new people
bleeding a calm sorrow. A
mistrial of hope we all smolder,
we all forget—I thought I was
something righteous, but I just
wanted to be right. A dilemma
of seasons, an endless night.

We believe it because we regret.
What we've done
we did on a promise.
A sustenance. A corpse blossom.

But not enough to end
our habits—
change our lives
our miseries.

A majestic site
to behold.
A majestic burnout
held within the soles

of the sentence, the fat sizzles
the skin, a misborn silence

we perform in mundane rituals.
Pray away our mystic order
to prevent casual faith–ash accreting

on our hands and feet and still

the acerbic silence, the raw decapitation
borne into the night.
This burning roar
this ignited hunger.

We are clear and sick and gasping–
a new kind of hush because we believe
in morning we bleed true,
we bleed swift, we die into lying

fungus. The morning marks
reticence–twirling leaves
in the distance.

I buy, sort, worry– like prayers

into the night–we must find our gristle
must right our undoing.

Clear woven ritual, camping
through demise we ply the heavens–
strip

the flesh off each new life booming.

Thieves dabble at my side.
Take all the good parts of my memory.

The land still, miscarried.

Stay like this with me forever. Your voice
hopeful, eyes tethered to stars.

What goes unnamed at the end.
What Armageddon do we exhale.

This excess you must skim
this new mantic passion.

Ripe for bias   this night asylum.

In the end, we compost
a new orison—

Don't forget me where you stand.
Don't forget the sorrow.

‡

I swallow my heart.

The day goes plastic.

‡

Our contagion
        flees into larks.
Our larks into
        intent and we intend
to mark our lot
        with flocks of empty
pockets we hold
        our scars within and sort
our larks into thin ferocity.

A courageous intent
        ignited another
intent and
        this time it was war.
Soon, the bodies
        dismembered our
memories and put
        pots into pans
and pans into solid locks
        like solidarity–made this
whole beating
        rancid, this whole
stain immovable
        a pillar of no.

This epoch among us,
    a thunder grip
narrowing its
    around me.

Another foot
    washed ashore.
Scorned my
    anxiety roars.

Mottled, I held
    my lot. Contained
sleet like sand,
    scurried my life's culmination,
and stuck trauma–hurried–
    where I could
try to save myself.

Detritus of every
    single day I parted
blessings, unmoored my silence
    and winnowed
what I could from my life.
    But we were already
beastly, sickly human
    scraps folded into nooks,
balled like inconsequential
    receipts. It's here
I turned mute.

Recorded every word
I learned like a recipe
suffused.

In my path,
a destruction unfolding.
Blasphemed, a new hope
shapes a melody.

What have we
become who do we
give it.

‡

Brackish acid beamed like electricity these humans wasted upon us. I renew my verdancy like a larch dropping its needled jettison, its promise to mend. I make solemn this tingle. That hope stored for private angles. We were never meant to last as long as we did. Were never meant to calm what nature ordered. Coming up for a blast of oxygen, the sun shatters against behemoth structures, over your patient sprawl, your acquired body, the ice melts from closed lashes like a listless cold kiss. The glow meant for every human. A desire to be clutched.

‡

gloom of inaction
        disease of reactive glands
infected toward the propaganda
        we waste a life
in regret

‡

deprived of that order

the slicking slice of erasure.

Forming a void where once your body lived beyond

itself–beyond an impossible echo.

What does it sound like when you extract

yourself, does the body burn? This is it, *the sea's frothing blue.*
A burly sigil. A swirl of ocean-clot flushed from our marrow.
A steamy bog of love and sorrow. And yet–

You said "

Broadcast your moment out."
It's enough to articulate

when you cannot speak–chance it all on black.

What I say is an ineffable hum. The tinge of sun demands the morning meet your eyes, we have traveled far and I am lucky to have been a passenger. There is so much power in the decaying resonance echoing into how words create this beautiful music. You taught me to extract our breath and how we become mouthy when electricity subverts the amplitude, listen to me yell, don't do it. Don't splatter, we'll become a waffling tone and that's it, you're sawtooth. You're grief plastered in our bones.

I can only thank you   for finding me on this comet.

That we have to be sentimental
when it matters. That I have to say it twice.

It's a rite and I know you   want to quietly fly.
But here's the problem   yelled into a void.
You're going to bounce off something.
We're going to make it loud, going to make it drone.

The laugh tracked into   persons not present   in situ
the burden imprinted   bare.   Seared into a memory.

Knocked into turgid king tide, a wave tumbles endlessly until it disperses, mimics on us simulacra of what has existed and will exist always. Love. Sound. Immensity. It's in the leaves today, it's that burning flack you've gutted into me. I know I need to share

the hole with others. Burden the
proof in how you've blown the
world away. Given to it, a score, a
remarkable panging note.

Behold the remedy, a crescendo in the moon light
within the corrosive skeptic's spasm.

You are always gravely aware. Home within
the art, a tree falling on nice things. Light among us.

How this line can never reach you
but I know you already know it. There will be silence

but what is new will awaken in us generations
malleable to your shrapnel. A harping bruise.

Churn hard turn that melody into your own

sound. This beat begins 4/4

and scatters infinitely into every possible heart.

Such is a star gone supernova.

We become paralyzed when existence
reflects back—that undoing

order of anything romanced from death.

Well within this world, it's never long
for this beautiful spectacle, this horror
outlasting.

Our orbit, our descent– a terminal
escape velocity.

Prayer in accumulation. Prayer of everything
we become. Prayer in mistaken singing.
It's all we've rehearsed.
It's all we return at dusk.

My friend, we'll find the experiments
to bring our nothingness together.

Purify the water into what matters
to mark ourselves with things
that persist. It's the lightning,
the flare of the sun toward earth.

The imperfect hidden scars.

Lain into the departing path
bodies become trajectories even if they exasperate

like a stool where you once sat crooked.
In this, our flock defines our unnatural
spasms in terms of lonely hymns—that which cannot
call back and demand to live again
even though we want it so badly.
The body grinds its microscopy faintly
into energy exploding
our silent, daily reproduction.

There are ways to listen, but it's never deep enough
to fill in the holes we dig, the bodies we sink at sea, and
the immolations we stoke. We breathe in.

Grief is inherently outward to counteract
the void within.

It's not enough to code language in metaphor—
you're here you've arisen
and like the dayglow gods you'll persist

because I insist that you do. It's not enough
to collapse the world inside out

vying to keep distance   between those who loved you

and those who just will never know—it's intrinsic that we prevail

and in our prevalence   relent, eventually so that we

can clasp the underarms to help

our survivors swim ashore.

To resist clouds, water masked in grey drops our dark into living unseen, we speak out, livid, the ones we love. It's impudent to ask, impudent to relax—to be left alone and let thunder roll over our songs. We'll never have enough rain. Never be able to sing the refrain. Spring will come anyway.

Writing won't save me. The sun will explode
and I'll shrivel into archival groans,   a knot tightening
to remember   its homunculus form.

Creation is not the question. Persistence
a form of resilience we lost

when you left.

It's never enough to stop trying
to remember that we can never
replicate your replicant, of memory a shattered mirror.

A fidget in the wind.

‡

Today, I had another seizure–
captured all of myself
in the small portions I divided
my brain into–I took illicit

the exasperations across the street
and threw them into the ocean,
imparted a turgid chill.

Every muscled spent.
Every body split.

I became the child I had always
wanted to become
– glacial, churned–
glancing toward

bigger delusions
and stopped

a frigid hope from setting
in. The cutting rain enthralls
my exaltation–
running through the squall
soaking in winter's snare

like strings
fraying
when the wind falters.
Over flappable creases, we blossom

memories into a kind of siphonophore–

prognosticate loose ends of our existence

into torpid, tornadoing bellows.
These myths we believe shape
calloused malice if we aren't careful on the dock–
a monstrous attitude toward failure.

What does silence hold?

In roars between waves
I carve my face effulgent–
all at once marking the emptiness
all around. We create loneliness
because we are born

an echo panging within ourselves.

A gumption of want, of existing near stations
beyond our centers. A beacon
to welcome or push away.

Painted history creased outwardly—
how cruel we must seek the pieces
and the tragedies we parlay.

Sometimes we're hostages.
Other times continents love travels over.

Pilgrims with knives stuck in our chests, a forgiveness
that yearning clears concisely within us.

A sharp panging hope.

The beach brays. I burrow. I rot
like a wet man. An idea
convulsed through years of insatiable waves.
A parting swallow until nothing remains.

‡

I used to believe in miracles.
Now I believe in the prophecy of the mortal sun.
The undying chasm of humanity.

‡

All water pulls toward the ocean, empties everything we pour into ourselves. For what? The brackish opening of everything that was pouring to an end—the void of possibility where we comb through a minuteness carried within the waves, a wanton throng washed ashore.

Falling from our prime to the zero-level nothing, the centrifugal wave razed away like a tidal march, pulling, gone, lazy coming freight towards grief's inner sight. Today, I found out my mother has cancer. Every birth becomes a weight we have to excise, a magical ruin and everything.

Hoarding our future from each other. Planting plastic amongst the plastic. Paying for our lives through a currency that doesn't demur. A minaret maneuvered into every last gaping wound.

Solemn prayer shouted from a ladder. Solemn prayer to inflict life. This pardon. We sauna to release our concrete permanence that never lasts.

Tell me to barge. Tell me to charge into silent inaction. The willful misery of inconvenienced lives. We stand there, watching, burning while the mountain shimmers fire, while the end glimmers. So much of life loses itself, ruminates remnant flotsam, paces the memorials into memories. Bronze hooves. Liking as a kind of solidarity, the fervor, the vigor, the passion pitted like a benign tumor bigger than your body.

—and then all at once, even though we saw it
close the distance, close time, we collapsed,

in each other, an echoing agony, that even so close,
every day, we were still, vastly misunderstood, appropriated.

Cowlick, on my face, you were everything you could be.
A lover to make music, from within, a caustic wing fluttering.

Here, now, hold me in your hands. Unfurl me slow,
like a minuet, crescendoing want, rustic indecision,

we scatter, we matter, more than ever, we become
two furious mornings, against a backdrop, where

we cannot even begin to believe, we are, motioned,
the lithe mystic, a callous between nights.

‡

These broken hallows.
          These veins pump at the earth's excess.

Blending distance with intimacy,
          brightness catalyzes reliquaries
          to take in new friends, new enemies.

Giving up ghostly
          passions, I resuscitate
          myself as wolves nip

                    at the threshold
                              where we begin–

We began to misshape the past
          began to blast
                    apart our bust–
          a kind history
                    to be forgotten.

In this silence, a glowing forge
          assembles good ballads

          that strangle out the fervor
                    that blissful malady.

The hope wrapped around
the moon's constant pull.
Always wanting more.
Always taking what
it can get.

‡

My teeth began hurting.
Began growing into each crevice
they craved. In this maw—
I mouth only that which sustains me.
Omit genuine inflections
that taper off into multiple kinds
of outward silence.
I'm afraid
I'm rotting from within.

Ask of yourself are you
prepared
to flail
the body flayed
itinerant.
We begin slowly.
We begin to breach
the whole heartbeat
bleating.
Strophed. Beached.

In the morning
I gnash.

In the morning
I streak.

‡

god in the emptiness

the inhuman

pang of nothing true

a tallow hope

too far from the fire

we believe

‡

We are a summation of our failures–
Like an exploding flower, lupine-craven, yarrow-golden, beauty overtaken. New murmured stallions. Riches smacking lips and then missions to other planets, consciousness mimicked into machines. As columbine wilts, it scatters gilt renewal. You're loved. You're not loved. You're a wholly misshapen art.

But tell me, how do you live beautifully in the bardo?
How do you mold spectacularly

in what binds us furtively
and bangs our mettle? Yesterday, I didn't risk enough

to fail, but is entering the world
enough? This harrowed plea, this mockery

still, a sadness borne from parts
of all the shame we carry. Indiscriminate earth,
slowly looping constant death–drive into this wall

I have shaped within me. I want to free
the mutant parts that plague me like a desert.
Screech a yawn

so uncomfortable I have to look away.
Fraught saguaro
of the everyday.
A somnambulist lullaby.

You're glistening
like a still breach
of shore.   A tidal
pool mired
in exploding
vibrant life
just before the sun
begins to linger
overhead.

‡

Sea stacks eroded
into quiet sentinels–

dregs of accumulating
rigid plastic colors.

‡

A voluminous clamor
sucking water from deep

stone-creviced pockets of waves–
a bridge unmade by time.

‡

Roped over logs, this detritus
shipwrecked like a snare–

knotted like an infinite
tangle–a broken shore left bare.

‡

Gravity, this unwieldy wretch, a sticky sea foam bereft.
And you, becoming cold, a breeze of all we left.

‡

Building dust upon the epoch streaked by the sun we learn we cannot preserve our memory.

‡

Reclamation of the void
Reclamation of the body
Reclamation of heartstrings detached
Reclamation toward less

It's that, or your life

We take a small moment and listen

The world wails
We want

a new sorrow
a new way to name it

this bleeding charter
this ruin wrapped in discarded parts
you've signed your name
you cried your eyes out
of their sockets

name your own price
name your fallow

It's an echo
no, it's the simulacra

prying open
a universe
you must acknowledge

Reclaim the word
Reclaim your life
Reclaim that mallow that made you sing
Reclaim the Sunday you laid naked your wife
your lover your husband your kink

Reclamation of tin cups tossed aside
Reclamation of needled wants
Reclamation of the polyester blouse
Reclamation of choking parts

Its clement music
The weathered wail
that makes a melody
Your voice and mine

a distance closing fast
like a garden churned

apart I sing this sorrow
I bleed hollow
that galloping wind.

Intertwined we stare down the ridge
It's in this harmony we ply

faith in each other. It's in this harmony we hum dearly. It's in this harmony we believe in the future. We're strong today but forgetting we're not someone new. Manifold pitch in the harmony just before complete cacophony–the frost squeals when the forest ravages an orange epiphany and ignites so much so quickly. The glass and twisted aluminum remain and maybe next year we might find morels.

Intertwined that bridge

between lives and living distant.

Burrowed

like a tragedy one of those stories that

exists between epochs–

it's you. It's me.

I thirst for your flesh.

Its memory.

The space you once existed

between my head.

The crushing shore.
Separating the seams of us.

Preservation of the void
Preservation of the body
Perseverance seeping night
It's a mischief mistaken
marred like the mystic
we lost in sight
I'm beginning to realize
I've forgotten your face, the way
your hands swept with it.

Reclaim want
Make reticent a nation
Reclaim the body
the corpus, the animus
the intent—blistering against
the boundaries, morning binds us in our sleep
we awaken our plight

Reclamation in plumbing
Reclamation in song
Reclamation as history singes
Reclamation in things we remember
without fear—

We lay it out so wholly
let the winds blow it all over

the wash over me wash over windows

singing this discord
made sweet– melodic
animal screeches.

A new partner to mimic.
To seem together.
A pattern like rain.

To the mountains
our feet pound
a line in dry earth. Exact

a menaced hope,
a calloused determination.
The salmon run
a stream beside us.

Hum silent cold
humanity. Hum
misery's unborn
plastic spasm. Our
unhandled world. Hum
because we want
more than we
can do.

Reclaim the silence. Reclaim
the frond. Reclaim the spinning

shoot. It's the feeling
when you open
your hands and
a handshaped shiver
rolls over your spine
in electric tinges,
a ventricle
pulsing itself alive.

Hum as the unbridling
Siphon our future hum
Calcify eons
we have wasted. Hum
the dalliance hum
the destruction we
wash in alkaline revelry.

Hum into the door—my enemy
unjars this wisdom, sluggishly
twists entry for the dead
Reclaim your apparatus
Reclaim the blue flickering
Hum our oeuvre chasms, together
this silence bares our misery portrait
My unanswered prayer

Perpetual fog
Perpetual gale
I'm failing to extrude

solicitude in this coastal awakening—
the trees claw at the mist
as it evaporates
try to hold the clouds
like an innocent surveyor of life.

Perpetual breeze
Perpetual fire

It's how we rise to boil.
How we fend off organs
fighting forever to make new this
body.
These human shores. These blustery
orgasms. These
entrapped bodies.

We radiate strong the sound of life
Reclaim the silent creepy kudzu limbs
this enraptured creeping Jenny.
Reclaim the violent yawp throughout history
Reclaim our place in this misgiving
Reclaim the fight from the tomb
of the stillborn bawl
I am hearing panting
I rub my feet to stay warm

And yet

Discovery makes new these dead friends
        Here we reclaim what we can
Unread past constellations guiding
unknowable destinies
Burbles making distance
        away from our hearts

Reclamation of the tide.
Reclamation in the wash, the ablution bubbling new each passing truth—
        sometimes we forget to miss her
Reclamation in the touch of each passerby
Reclamation in assuring their weight, their marrow.

                Back home
                        the soil is always wet
            always blackening toward
                possible life. In the mountains

        we cake dry
            regret and all the bones.
            We stream our grief

            in cool deluge to calm
                our ambitions our
                hallowed youths.

In the morning we'll reclaim our quiet rage. Our world changing

moments—we don't get to be comfortable.
We don't get to decide to step away.
It's greater than us.
Greater than indecision.

Defining our collective witness
like a tree
stuck sideways—starting

to drop the benevolent
belief, this human reclamation.
This seed planted queer—
holding dear what we ought
to fear.
This thousand-year stoic.

The deafening roar
of fire, a giant
ocean of trees burning
like a sunset.

‡

Fondle this masterpiece
this new coastal glide.

‡

How do we face
extinction and forego

the bodies?

Most of all
the bestial
the insectile

a fertile nullification of everything raw.

Supine death pose,   a cathedral
of limbs.

And instinctually–we witness, we pile
our prayers toward the infinite,
the finite erects an accumulation

with distance and memories.
The barren emptiness
within every one of us waiting
to get out.

‡

To be an arsonist

when the West

tinders.

How menaced

we find solace.

How alone

we smolder.

‡

When I want to die
in myself

loosen my human frame

I crawl into the desert
to witness the infinite

whole of sand–
the earth crumbling

stardust to stardust

epic to epoch

sentiment to sediment

a dance words twirl
as they echo into the abyss.

‡

The future opened. Ignited right

in my eyes. A corpuscular
popsicle. The moon
gone

red. An adumbration

swooning night.

‡

Find the face of another who loves your face
and the boiling viscera beneath it. Make this a home.
Be as transient as you need. Within other
people there is a mystery upwelling,
dispelling like an inhalation—as fleeting as
putting on a warm sweater, we recycle our prayers.

Make new our gods, our lost salvation.

I had faith in people
to let me down
off this self-inflicted journey.

Here's the truth: The sun visits
        less often.

Sand will weather away
that which will wither anyway.

Smooth me into rupture,
stroke microscopically my inflammation—
this layered madness a tinged festoon.

        Place my thumb
                on your tongue.
                Like snow forced down

our backs, we recoil
we snipe, we fierce.

Place this, my whippoorwill, down my throat, let it speed
away.

Jettisoned, ready, and immaculate–my heresy.
We were put here to suffer
the words until we couldn't
stand and say them.

‡

I miss you she sighs before the plight she sighs a pitch she sighs pleated ocean my boundaried vision rising she sighs then she sings. Before the fire burns she sparks a sweet ignition.

And you
extracting the golem
we left together.

A sign of hope. A promise
of possibility.
We let

the others follow
our footfall.

‡

I succumb to the hunt
of all things.

Endnotes

This book's title is derived from John Keats' "Bright Star" sonnet

Page 13: "like a body wholly body" is from Wallace Stevens' poem "The Idea of Order at Key West"

Page 30: "breathe out enough/to breathe in/and take another life" is inspired by Michael Dumanis' poem, "The Woods Are Burning" from *My Soviet Union*

Page 41: This poem is for Kirsten Rue

Page 46: "we cheer our awfulness we've prayed" is a misheard St. Vincent lyric

Page 53: "you must change your strife" is inspired by Rilke's poem "Archaic Torso of Apollo"

Page 54: "we have shadows/to earn" is inspired by Ben Lerner's line "We have birthmarks to earn" from *Mean Free Path*

Page 67: This poem is made up of a litany of headlines about climate change

Page 69: "My enemy" poems are inspired by Jane Gregory's book *My Enemies*

Page 102: The series here is inspired by Lisa Ciccarello's book *At Night*

Page 112: "or else we swoon to death" is from John Keats' "Bright Star" sonnet

Page 120: This poem is for Julia Madsen and has a line inspired by her book *THE BONEYARD, THE BIRTH MANUAL, A BURIAL: INVESTIGATIONS INTO THE HEARTLAND*

Page 120: This poem has a reference to Jake Syersak's book *Mantic Compost*

Page 130: This poem is for Prageeta Sharma

Page 148: This poem was inspired by an interview with Elizabeth Holmes

Referentials

Thank you to the editors of *Afternoon Visitor, Black Sun Lit, The Destroyer, DIAGRAM, Gasher, The Georgia Review, Heavy Feather Review, mercury firs, Paperbark, Poetry Northwest, Provincetown Arts, Tyger Quarterly, TYPO*, and *VERDANT* where many of these poems first appeared.

Many thanks to Quenton Baker, Bill Carty, Lisa Ciccarello, Sean Cleary, Carl Corder, Justin Cox, Stephen Danos, John Englehardt, Jade Gee, Luan Heywood, Paul Hlava Ceballos, Burke Jam, Torin Jensen, Lars Garvey Laing-Peterson, Julia Madsen, Matthew Oglesby, Caryl Pagel, Alyssa Perry, Juan Carlos Reyes, Alison Riley, Nicole Rue & George Resor, Philip Schaefer, Matthew Schnirman, Prageeta Sharma & Mike Stussy, Jake Syersak, Jeff Whitney, Jane Wong, and Deborah Woodard. Special thanks to Dan D'Angelo and Ryo Yamaguchi. Much admiration to Johannes Göranssen and Nick Gulig for your kind words.

Thank you to places that provided the space for this book to drift into being, including Draft Punk, Flying Bicycle Cooperative, Ocean Haven, Stone Way Cafe, Tangletown Public House, and Third Place Books Ravenna.

Thank you to the musicians who provided the soundtrack to writing and editing this book, including Ben Frost, Boy Harsher, Broken Water, Daniel Bjarnason, Edzayawa,

ENOLA, Fela Kuti, Jóhann Jóhannsson, John Grant, Jonathan Bree, Lala Lala, Marlon Williams, Mating Ritual, MOD CON, Mount Eerie, Princess Chelsea, Tropical Fuck Storm, and Yves Tumor.

Many thanks to the UNESCO Cities of Literature network for all of the inspiration and constant reminder that the global literary community can be smaller and better when we work together across borders.

Immense gratitude and reverence for my publishers Carrie Olivia Adams and Janaka Stucky. Black Ocean is a special endeavor that continually inspires, challenges, and makes the world feel simultaneously more expansive and less lonely.

About the Author

Josh Fomon's first book, *Though We Bled Meticulously*, was published by Black Ocean in 2016. His poems have appeared in a variety of journals, including: *Afternoon Visitor, Black Sun Lit's Vestiges, Caketrain, DIAGRAM, The Destroyer, DREGINALD, The Georgia Review, jubilat, mercury firs, Paperbark, Poetry Northwest, TYPO*, and *Yalobusha Review*. Josh lives on the unceded lands of the Coast Salish peoples in Seattle. He can be found at https://www.joshfomon.com